Lifespan

By Robert J. Alderman

Preface

100 years ago people would laugh at you if you said "Someday they will be able to open your chest and replace your heart with a mechanical one. They will even be able to successfully take someone else's heart and transplant it into your body." Likewise if you said, "A microscopic DNA molecule can store all of the information necessary to make a complete human being."

Dr. Benjamin Carson has successfully taken out one half of the brain, in children, and they then lived a successful life. Wikipedia: "Carson's other surgical innovations have included the first intrauterine procedure <u>to relieve pressure on the brain</u> of a <u>hydrocephalic</u> fetal twin, and a <u>hemispherectomy</u>, in which a young girl suffering from uncontrollable seizures had one-half of her brain removed." This is common knowledge today.

The "Discovery"

This discovery would forever alter humanity and it happened in 2009. At first the impact that it would have on the world was not clear. It was only after several years of experimentation that the nature of the discovery was fully understood. This discovery was related to the original question "How can a complex molecule like DNA remember how to make something as complex as a human being?"

Completed in 2003, the Human Genome Project (HGP) was a 13-year project coordinated by the U.S. Department of Energy and the National Institutes of Health. After scientist completed the human genome project it was well known that the DNA molecule had all of the information necessary to make something as complex as a human being and more. What was yet to be discovered was mind-boggling and changed the world. It was heralded as the greatest discovery of all time.

It all began in a lab in Germany, doing research on brain damage, spinal injuries, and stem cell research. They made a startling discovery. They found that small brain cell transplants were possible if certain DNA matches were made first. Not only did the newly transplanted brain cells take hold and live, they rapidly reconnected to synapses and functioned quite well.

In one case a man awakened from 4 years of coma within 72 hours after surgery! In this case human growth hormone (HGH) had been used to accelerate the brain cell synapse reconnections. The spinal cord research was encouraged by these results but so far no comparable success.

Gustaf Gottlieb, GG for short, worked at the research lab. His beloved dog, named Grover suddenly died. He had taught Grover many tricks, Grover could sit, fetch, play dead, shake, speak (bark), roll over, jump through hoops, and so on. GG was devastated by Grover's death. GG had recently been researching the part of the human brain responsible for memory. He had discovered a strange looking group of brain cells located at the junction of the Temporal Lobe which is associated with perception and recognition of auditory stimuli, memory, and speech, and the Occipital Lobe, associated with visual processing. These strange brain cells were more dense and smaller in size than the surrounding brain cells. In a moment of shear genius, GG decided to transplant similar brain cells from his dead dog Grover's brain to another living dog. He ran down to the dog pound and got a dog named Boo. Boo was to be euthanized later that day at the dog pound.

The Operation

The first thing was to surgically remove only Boo's brain cells that, GG hoped, held memories. This group of cells was only about the size of a peanut. These brain cells were removed then replaced with the similar brain cells from Grover's brain. The operation went smoothly and Boo was up and about in two days. Only about a week had gone by when GG noticed that Boo was acting like he had known him forever! He would jump around and get excited every time GG would be in the same room. GG would call him Boo and hardly get any response but when he called him his dead dog's name, Grover, Boo would go crazy! GG decided to treat Boo just like he did Grover, when Grover was alive. He put him through all of his tricks. Lo and behold Boo knew all of the tricks that had been taught to Grover. For all practical purposes Boo had mentally become Grover!

Excitement at the lab

A meeting of the scientist at the lab was held for the purpose of presenting GG's work. Of course on everyone's mind, could this be done with humans? It was decided to schedule an experimental surgery with a 25-year-old man, named Al Gochen. As far as anyone knew he had no living relatives. Al had no memory capability at all. He was comatose in a persistent vegetative state. By law Mr. Gochen would be un-plugged from life support in just a few more days. His brain damage had been due to a self-inflicted head injury. In an attempted suicide he had shot himself in the head. The part of his brain that was missing was the same part that GG had associated with memory storage. The reasoning was that the small amount of brain cells to be transplanted could do no harm. The brain cells to be transplanted would come from a man visiting Germany named Walter Crosby that had recently died from a heart attack. Walter had donated his body to science and was frozen. Walter had no living relatives and had made arrangements with Bio Gift. Bio Gift asked the German lab if they would like to keep the body for experimental purposes? The lab gladly accepted the offer and Bio Gift was spared the expense of shipping the body back to the USA.

A vote was held at the lab meeting and it was unanimous. Since it could do no harm, we should try it.
GG immediately set up the operation to be completed in the OR at the local hospital. The hospital was small but staffed by well-trained doctors. They had cooperated with the lab several times before on spinal cord experimental surgeries. GG brought Walter's memory brain cell specimen with him in a small-sterilized thermos jar. The brain cell specimen was only about the size of a golf ball. The jar contained a liquid known only to GG. GG said, "The liquid is to prevent oxidation of the tissue." The liquid was actually brain fluid obtained from an

aborted, three month old fetus with added human growth hormone (HGH). The reasoning was that this special fetus brain chemistry was conducive to developing brain cells. GG had discovered that this liquid along with human growth hormone facilitates the connection of the donor brain cells to the existing ones.

Back from the dead

About a week after the surgery a nurse came running out of Al's hospital room screaming. Mr. Gochen has gone crazy! Imagine the shock and delight as Mr. Gochen said, "Why in hell are you people calling me Al Gochen? My name is Walter Crosby!

Troubles

The next few months were nightmarish for Walter. First thing was getting used to the new body. Walter was 70 when he died.

Twenty five year old Al Gochen's body was handsome, in good physical shape, had perfect vision, and was not overweight. The gunshot wound destroyed Al's memories; but everything else worked fine. It was much better than Walter's old body. Two weeks had gone by with constant adjustments being made by Walter. It was a small town and it seemed as everywhere he went some new problem appeared. Total strangers, to Walter, would come up to him thinking that he was Al Gochen. They would say something like, Hi Al, Ich dachte, sie seien tot! , or in English, Hi Al, I thought you were dead! Rather than say he was not Al, Walter remembering his German, would say, Al ist tot ich bin sein Bruder. (Al is dead I am his brother.) This was easier than telling the truth which was unbelievable.

Walter was asked to stay in Germany to let the lab study him for a while, all expenses paid plus $1000.00 per week. Walter had no money because after he died his will had dispersed all of his wealth to charity. He really had no choice but to comply. Each week thereafter a meeting was held at the lab to review and discuss this enormous breakthrough. When the German government got wind of this they dispatched a legal team to the lab. They made everyone, including the person they called Al Gochen, sign a non-disclosure, secrecy agreement for national security purposes. The fear was that if the knowledge of real, proven, life after death got out, panic and riots could occur. It could be that Germany was also sensitive to being accused of playing God.

As soon as Walter got enough money saved up he thought he would go back to Florida where he lived. The officials said no, that he was a German citizen (Al Gochen) and he was not free to leave! They used the excuse that he had been served with legal documents that accused him of owing back taxes; fines, speeding tickets and that debtors were also after him. Walter thought, "No wonder, this poor sap committed suicide."

Walter found that he had a serious legal problem. Who was he legally! Walter insisted that he was Walter and nobody else! But his face, voice, appearance, fingerprints, etc. were Al Gochen. He decided that he might as well pretend to be Al Gochen and try to get a passport to travel. He got a local attorney to file all the necessary papers and petition the court to settle all charges, pay the back taxes, fines, and then file "Insolvenzordnung" or in english "bankruptcy". This took a while but he was finally successful.

Home at Last

Walter finally made it back to Ruskin Florida as Al Gochen a German tourist. He was tired and bewildered but glad to be alive and was experiencing the most profound event in human history, true life after death. He wondered if he was really bound to the German government secrecy agreement by US law. He did not feel morally responsible, as he was never asked to participate in this experiment in the first place. He decided that it would be wise to keep this as quite as possible, anyway, who would believe him?

Walter knew that he needed a good lawyer so he called on an old friend who was also a federal judge. The judge's name was Henry Hudson "Chief United States District Judge, United States District Court, Middle District of Florida". Henry and Walter were close and attended East Bay High School together. When Walter walked into Henry's office the receptionist asked him "Can I help you?" Walter said, "Yes I am an old high school friend of Henry's. Tell him Walter Crosby is here to see him." When the receptionist buzzed to tell Henry, he came flying through the door and said, "What is going on here, Walter died a year ago. Who are you trying to fool? You sir, are an imposter!"

Well Walter had to calm him down before he would even listen to the fantastic story that was about to be told. Walter thought, "What can I say to convince him that I am me even though I don't look like me?

My face, passport and fingerprints say that I am Al Gochen. What to do, what to do." Then Walter remembered that he and Henry had done things together that no one else would know about. Walter told him about fishing together, rabbit hunting, and could answer every question that Henry could think of to ask him. In fact every time Henry would try to catch Walter with a trick question, Walter would laugh and point out that it was just a trick question. Walter finally asks the judge to sign a non-disclosure form and showed him the ones he had with the lab and with the German Government. This cinched the deal and the judge realized that it could be no other than Walter Crosby back from the dead! Walter and Henry spent most of the day reminiscing and trying to figure out what to do. As Walter was leaving the judge said "Let me think on it and I will call you in a couple of days."

Money fears

Financially Walter was down to his last $843 dollars. He was very worried that he would not find a job under the Obama train wreck of an economy. While he was having a coffee at a wireless hotspot Walter decided to try and sign on to his old secret E-Trade account. All of his other accounts listed in his will had been closed. Just for fun, he nervously typed in his user name and password. Unbelievably, his account came up. It was showing over 1.7 million dollars in his brokerage account. He was ecstatic with joy. Up until now he thought he was penniless and had been very depressed. Never in his wildest dreams did he think that this account would go un-noticed. Furthermore his stocks had gone up beyond where he would probably have sold out. It must have been that he just forgot to include this account when originally listing all of his assets on his will. "Not bad for a dead guy, Walter thought." "Now how can I get cash out without arousing suspicion? What if the probate court has a hold on the account? Well I may look like Al Gochen but by damn I am Walter Crosby and this is MY MONEY!"

Walter thought "I could set up another account if the local bank will let a German national open one? Then I could transfer money to that account and cash checks on it? Maybe have a credit card sent to my motel address? I could transfer money into a friends account and then have them give me in cash but if this turns out to be a crime it may implicate them? Oh well I guess I will try the credit/debit card for now." It worked. "Later on Maybe a Cayman bank account would be in order?"

Two days later Walter realized that he could have already transferred funds to his bank in Germany and used his Al

Gochen account. Walter thought, "This will work to isolate my funds from any future unjust claims against me as Al Gochen in Germany. I will keep the German account low in assets, just enough to get by until I get this mess settled."

Good News?

Now that Walter had a nice supply of money he moved into a nice condo at Bahia Beach in Ruskin, Florida. There he soon got a call from Henry his friend the judge. Henry said "You are not going to believe this but we have an appointment with the US Attorney General in Washington DC."

It had turned out that our CIA already knew about the successful brain cell transplants in the German lab! "They want to meet with you in person." Now Walter was REALLY worried!

The next two days was rush, rush to get ready for the meeting. The judge was very helpful and made all the travel arrangements. He seemed very excited about going to Washington to the meeting. Little did they know that the President was going to be at this meeting!

As Walter and the judge walked into the office of the AG they were met by the AG and the President himself. This made Walter very nervous but he knew he had to get his identity made right. Never in the history of mankind had this type of problem existed. One thing for sure, his was the first, but millions more would surely follow. After the usual handshakes and amenities, the AG said, "The CIA has found out that 43 other transplants soon followed Walter's. All have been totally successful. Both the USA and Germany have agreed to keep this secret." It was believed that if this got out to the public there could be chaos. Life after death is a powerful desire. Now that it has actually been achieved and proven, who can predict the effect on humanity? Just look at what the un-proven promise of life after death has done. Tens of millions of religious followers have killed each other. Many have believed that they were going to heaven. Remember, Jim Jones Kool-Aid, Heaven's Gate Hale-Bopp comet, Muslim's 72 Virgins, streets of gold, Crusades, etc? This was all caused by just a promise. What we have done is real and provable.

The President blurted out, "Let's get to the reason for this special meeting! Mr. Crosby, do you know the secret to making the brain cells re-connect to the other brain cells? Dr. Gotleib has not told anyone how to prepare the brain tissue before the implant. Our scientists have tried to duplicate his work, but to no avail. We can only make it work about 50% of the time. That is not acceptable. Dr. Gotleib has chosen not to patent his procedure and instead he is keeping it a trade secret." Walter thought, "Poor Dr. Gotleib, looks like a case for water-boarding."

The president said, "We have already set up a secret Senate committee to make rules and regulations to govern this amazing breakthrough. After debate, behind closed doors of course, these will become laws for life after death services. These laws will eventually take the place of religions that can only *promise* life after death. Each person will have a legal record on file with the government. To be eligible for a memory brain cell transplant a person must have a very good record. Even the slightest criminal record and you will not get a new lifespan. Once the public is aware, they won't even jaywalk. Crime will drop dramatically. Police will mainly be there to help non-criminals. This is the most important medical breakthrough in the history of mankind!"

The Attorney General spoke up and said, "Whoever controls true life after death will easily rule the masses! Controlling memory brain cell transplants are what we in the Federal Government are after."

Now Walter was more worried than ever before. He saw himself being used for reverse engineering. Walter thought, "Next they will drug me and I will be on some ass-hold's operating table being dissected!"

This meeting did have one positive outcome for Walter. Judge Henry suggested that the President make Al Gochen an American citizen by executive order. The president wanting to keep Walter happy, and on his side, agreed. Passport and documents will be sent to his Ruskin, Florida condo. Walter walked out as a legal American citizen named Al Gochen. Walter only had to swear to keep secret his true situation, under penalty of high treason. Walter already had come to the conclusion that secrecy was the best path for him to follow.

Back Home

Now that Walter was back at home and settled in he decided to lookup a few old friends so that he could re-establish relationships. One that he thought of was another high school buddy, Joe Alderman. He and Joe had conversed many times before on a high scientific level and Walter thought that if anyone would appreciate what had happened to him, it would be Joe. Joe was also an inventor with over 110 US patents and knew how to keep secrets. It was hard to think of just how he could approach Joe without spooking him. The idea came, "Hmm I will just say I am Al Gochen, Walter's very, very close friend from

Germany and that Walter asked me to contact Joe when I was in Florida and that I was doing something for Walter that involved Joe." It worked, and Joe agreed to meet.

Joe came to Ruskin in just a few days and then the challenge was how to break the news of the transplant. It turned out that it really was not that hard, I simply followed the same pattern as with Judge Henry. It did not take long after Joe had totally accepted my situation that we, as usual, ended up with a discussion of Joe's latest scientific paper on "Time". The following is the paper:

Why is it Always Now?
By Joe Alderman

First let us think about how the concept of Time could have started. Somewhere along the evolutionary trail early humans developed memory capability and also imagination. Memories created

the "notion" of a "Past" and Imagination created the "notion" of a "Future".
You can remember a past or imagine a future. You may even dream of a past or a future. This paper attempts to prove that reality and consciousness only exist in the present. The past and the future do not exist. It is always now.

How Mankind Got Off on the Wrong Foot With the Concept of TIME.

When the mind developed within our human brain it did so through evolutionary pressures. The mental development was steered by whatever favored survival of our species. Having a memory was very beneficial.

Memories played a large role in creating our primitive concept of TIME. We can remember sequential events and this mental process suggest a past. Memories are more like a movie film recording of a certain sequence of events. They did happen and in a certain order but do not still exist as a "PAST". Cave drawings indicate that we have had memories for a large part of human existence. The cave drawings are evidence that cavemen had a memory and could visualize. Evolution favored those that could remember things. Like where the food was, or "WHEN" the salmon ran, or to prepare for the cold weather that will surely come.

  Being able to imagine was also a powerful advantage. Imagination created the notion of the "future".

"Clocks"

Actually, time does not exist independently. What we refer to, as "Time" is the concept that describes the relative motion and changes of matter. Generally it is the constantly changing arrangement of all matter that we are referring to. <u>All clocks ever conceived simply compare one motion to another.</u> Relative motion is what is actually being measured. It could be the motion of grains of sand in an hourglass, of water in a water clock, regular pendulum swings, atomic vibrations, Earth and Moon orbits, and so on. All "Time" measurements are really only comparisons of relative motions. The word Time represents the totality of these ever changing relative positions, directions, and relative motions of matter.

What we call the Past is nothing more than the "arrangements "of all of the matter in the Universe that preceded the present "arrangement". It is the constantly changing "arrangement" of matter that we call Time.

PROOF

Try this simple thought experiment. Imagine if <u>all relative motion stopped</u> even down to the electrons within their orbits. What we call "Time" would also stop. The current "arrangement" of all matter would stay the same. The hands on the clock of course stop moving at the position they were in when the freeze took place. One can easily see that without the relative motions of matter, time does not exist and has no meaning. This is because fundamentally it is the relative motions that we are describing when we use the term "Time". This is one simple proof of what "Time" represents. Time does not exist independently. Time represents the "*changes* in the arrangements" that are taking place. Time is only a conceptual tool used to relate the complex, changing, positions, directions, and motions of matter to other pieces of matter throughout the Universe. The hands on the clock are moving relative to all the other motions in the Universe. This motion relative to other motion spawns a circularity.

Without matter to compare to other matter, distance is meaningless. An example of this is when we "measure" distance. The devices we use to measure distances all simply compare at least two pieces of matter in some way. If all "relative" motion stopped a three-axis coordinate system would locate every piece of matter within space. Matter is moving about so another factor is needed to describe the "change" in relative positions of objects within space. This fourth dimension is what we have been calling TIME. We have been using this concept of time for so long that it has become part of our language and thinking. This has nearly <u>blinded many</u> and has caused most to think that a Past and a Future really exist contemporaneously with "Now". <u>The Past and the Future can not co-exist. It is always "Now"</u>. Time is merely an expedient way to express (and think) about the very complex relative motions of matter through space. The word <u>Time should</u> represent the changing relative positions, directions, and relative velocities of matter within space. Simply put, "The Arrangements".

NOW

When someone asks, "What is the time?" You can never answer with absolute accuracy. Even as you were saying the words, relative motions of unimaginable complexity have already occurred. Even the hands on the clock moved ever so slightly. The Cesium atomic clock in Boulder Colorado vibrated 9,192,631,770 times each second as you spoke. You may have realized that "Now" is infinitesimally small. No matter how small the increment, some relative motion of matter has occurred somewhere in the Universe. "Now" is fluid and constantly changing. Only the current *"instantaneous arrangement"* represents "Now".

THE PAST

If we think about the past, memories and imaginations are conjured up about some mysterious time gone by. The reality is that we are referring to an infinite set of arrangements of all of the pieces of matter in the Universe. Time travel into the "Past" would require that all of the intricate pieces of matter in the Universe be returned to a previous arrangement of the relative positions, directions, and velocities that existed at that "instant". This is ludicrous and impossible. The past is probably thought to exist because we have "memories" stored within our brains. These are only memories and are like photographs. It would take an <u>infinite amount of matter</u> for even a small amount of the past to exist contemporaneously. Imagine; you would need a complete Universe for each microsecond of Time. Consider two time travelers at two different times in the past. This would mean that at least two "extra" Universes would be required, one for each of the two time travelers. Ad infinitum! The "Past" does not exist, but historical records and memories of past "arrangements" make it seem so.

THE FUTURE

This is the same infinite amount of matter argument as the Past above. The future is simply a reference to the relative arrangements, positions, directions, and velocities of all the matter in the Universe at a step that has not occurred yet but will. The "Future" cannot exist contemporarily.

TIME TRAVEL

It is truly laughable to hear educated, seemingly wise men discussing with all seriousness, time travel through "wormholes" etc. There is no "Past" or " Future", in independent form but only ongoing changes in the "arrangement". In a sense the dinosaurs are still here, but they have "changed", their particles are just not assembled in the same way. They have long since disintegrated and their particles have been rearranged. Some dinosaur particles may even be contained within our bodies "Now".

Parallel Universe

By definition "Universe" means:
<u>The Universe is defined as the totality of everything that exist.</u>

The Greeks created Atlas as the Titan giant whose job was to hold up the World. They could not imagine the World without something to hold it up. Atlas of course was not really needed to explain anything. They just made him up because of their limited thinking ability and lack of understanding. When some folks can't solve a mystery they invent things to make their inept theories work.

For parallel universes, the past, or the future, to exist, they would be infinite, in number. This would take an infinite amount of matter. This is absurd and the gravity fields generated by this infinite amount of matter would also be infinite. There can be no past, no future, nor a parallel universe. There is only one universe and it only exists NOW. It is the constant "re-arrangements" of matter through motion, that we call Time.

AGE

Age is another word that really means <u>change.</u> The word "Age" has been used so broadly that it has taken on an independent character. There is no absolute "Age". Age should refer to the <u>changes</u> of the particles that make up the object under discussion.

You have heard that one thing can age faster than another. People, wine, etc, this is just an example of the widely perverted use of a concept. What people are really referring to are different "relative" amounts of change. It is the different rates of change that we refer to as different rates of aging. A sharper more disciplined use of such terms should be required for scientific thinking. We sometimes let our words blur our thoughts and our minds eye vision, to near blindness. Memory and Imagination may be the mother of Time?

To sum up:
The Past, and Future, cannot exist contemporarily or in independent forms.

Time represents the constantly changing and rearranging of the relative positions, directions, and relative velocities of matter within space. Time does not exist as an independent variable. It is simply always "Now". "Now" is the "instantaneous arrangement" of everything in the Universe.

Einstein wrote about time dilation believing that light traveled at a constant speed of about 186,300 miles per second. This creates many problems. Another way to explain the constant speed of light is that it is a "receiving" phenomenon. That would require a field of some kind that exists throughout the Universe. It could be that the existing electromagnetic fields from all matter "fills" the Universe and light is a vibration of this "collection" of fields. This would make the velocity of light a constant regardless of the source.

The End
Of Time

Joe and I discussed this paper until 2Am and finished off nearly a whole bottle of Dewars scotch. Joe eventually went to sleep on the couch; he was just too drunk to drive.

Next Day

Amazingly when I awoke in the morning I was convinced that he had to be right. It is always "Now" and what we refer to, as "Time" is nothing more then the continuous re-arrangement of matter throughout space. The obvious observation, that it, would take an "infinite" amount of additional matter for a Past or Future to exist, totally convinced me. Joe's argument does not require an "invented" dimension or made up factors to make sense. The paper also proves that when we say we are measuring "Time" we are really only measuring the motion of one thing relative to some other motion of some other thing. The "thought experiment" (Proof) made this very clear in my mind. Stop all motion and "Time" stands still. Obviously Time did not exist independently in the first place. Time is relative to motion and only represents the constantly changing "Arrangement" of the Universe.

Joe gave me some credit for suggesting that evolution would have certainly rewarded early man as they developed a sense of Time. The caveman had both memory and imagination capabilities. Witness the drawings on the cave walls. It all fits with Joe's theory. I imagine Man the hunter or lover, thinking that if he went to a certain place at a certain TIME he would have an advantage. All of this was really based on his memories and imagination. The diurnal cycle of the Sun coming up and going down, falling asleep and awakening each day new, certainly had an evolutionary impact on early man. We have that powerful indoctrination engrained in our brain today!

Early Man's "memories" and "imagination" may have given birth to the concept of Time, *Memory* creating the concept of the *Past* and *Imagination* creating the concept of the *Future*. My brain cell transplant of the whole set of my memories are what I attribute to what I call "ME, Myself and I" so memory and imagination may also play a part in consciousness? When Al Gochen blew out the brain cells that stored his memories, he also lost consciousness. For all practical purposes "He" was dead. The abnormal mixing of memory and imagination may also play a part in some forms of mental illness.

Walter proved who he really was beyond any doubt when Joe awakened to a nice breakfast of Mullet roe and grits, Joe's favorite. Joe always referred to the meal as "caviar and grits" since roe are fish eggs. Walter had once been the Mullet king of Ruskin. Now that he was in a 25 year old body he could throw a 12-foot cast net with no problem! Hail the King! The King is back! Joe was 72 now and he could not handle those big heavy cast nets like he did when he was young. Walter had a good supply of Sand Perch, Mullet fillets, and even a few Sheep-head, in his freezer. At night he could walk along the seawall near his condo and catch fresh mullet nearly any time. Sometimes an illegal Snook would get in the net and Walter, being Walter, would filet and release it, He called those striped Mullet.

During the breakfast conversation some far out ideas were presented. One of particular interest was "What if you could clone yourself from one of your cells and later transplant the memory section only of your brain into the young clone.

What would happen as the new, you, matured? Would the clone's blank memory cells dominate and erase your old transplanted memories? Maybe early on you could surgically remove the un-recorded on, blank memory cells of the clone and simply replace them with yours? This is what had been done inadvertently when Al Gochen shot himself.

Walter had been doing research on cloning and showed Joe the following newspaper story.

Actual newspaper story, April 22, 2009.

"Fertility expert: 'I can clone a human being'

Controversial doctor filmed creating embryos before injecting them into wombs of women wanting cloned babies
By Steve Connor , Science Editor

"A controversial fertility doctor claimed yesterday to have cloned 14 human embryos and transferred 11 of them into the wombs of four women who had been prepared to give birth to cloned babies.

The cloning was recorded by an independent documentary film-maker who has testified to *The Independent* that the cloning had taken place and that the women were genuinely hoping to become pregnant with the first cloned embryos specifically created for the purposes of human reproduction.

Panayiotis Zavos has broken the ultimate taboo of transferring cloned embryos into the human womb, a procedure that is a criminal offence in Britain and illegal in many other countries. He carried out the work at a secret laboratory, probably located in the Middle East where there is no cloning ban. Dr Zavos, a naturalised American, also has fertility clinics in Kentucky and Cyprus, where he was born. His patients – three married couples and a single woman – came from Britain, the United States and an unspecified country in the Middle East.

None of the embryo transfers led to a viable pregnancy but Dr Zavos said yesterday that this was just the "first chapter" in his ongoing and serious attempts at producing a baby cloned from the skin cells of its "parent".

"There is absolutely no doubt about it, and I may not be the one that does it, but the cloned child is coming. There is absolutely no way that it will not happen," Dr Zavos said in an interview yesterday with The Independent.

"If we intensify our efforts we can have a cloned baby within a year or two, but I don't know whether we can intensify our efforts to that extent. We're not really under pressure to deliver a cloned baby to this world. What we are under pressure to do is to deliver a cloned baby that is a healthy one," he said.

His claims are certain to be denounced by mainstream fertility scientists who in 2004 tried to gag Dr Zavos by imploring the British media not to give him the oxygen of publicity without him providing evidence to back up his statements. Despite a lower profile over the past five years, scores of couples have now approached Dr Zavos hoping that he will help them to overcome their infertility by using the same cloning technique that was used to create Dolly the sheep in 1996.

"I get enquiries every day. To date we have had over 100 enquiries and every enquiry is serious. The criteria is that they have to consider human reproductive cloning as the only option available to them after they have exhausted everything else," Dr Zavos said.

"We are not interested in cloning the Michael Jordans and the Michael

Jacksons of this world. The rich and the famous don't participate in this."

It took 277 attempts to create Dolly but since then the cloning procedure in animals has been refined and it has now become more efficient, although most experts in the field believe that it is still too dangerous to be allowed as a form of human fertility treatment. Dr Zavos dismissed these fears saying that many of the problems related to animal cloning – such as congenital defects and oversized offspring – have been minimised.

"In the future, when we get serious about executing things correctly, this thing will be very easy to do," he said. "If we find out that this technique does not work, I don't intend to step on dead bodies to achieve something because I don't have that kind of ambition. My ambition is to help people."

Dr Zavos also revealed that he has produced cloned embryos of three dead people, including a 10-year-old child called Cady, who died in a car crash. He did so after being asked by grieving relatives if he could create biological clones of their loved ones.

Dr Zavos fused cells taken from these corpses not with human eggs but with eggs taken from cows that had their own genetic material removed. He did this to create a human-animal hybrid "model" that would allow him to study the cloning procedure.

Dr Zavos emphasised that it was never his intention to transfer any of these hybrid embryos into the wombs of women, despite Cady's mother saying she would sanction this if there was any hope of her child's clone being born.

"I would not transfer those embryos. We

never did this in order to transfer those embryos," Dr Zavos said. "The hybrid model is the thing that saved us. It's a model for us to learn. First you develop a model and then you go on to the target. We did not want to experiment on human embryos, which is why we developed the hybrid model."

Dr Zavos is collaborating with Karl Illmensee, who has a long track record in cloning experiments dating back to pioneering studies in the early 1980s. They are about to recruit 10 younger couples in need of fertility treatment for the next chapter in his attempts at producing cloned babies.

"I think we know why we did not have a pregnancy," said Dr Zavos. "I think that the circumstances were not as ideal as we'd like them to be. We've done the four couples so far under the kind of limitations that we were working under.

"We think we know why those four transfers didn't take. I think with better subjects – and there are hundreds of people out there who want to do this – if we choose 10 couples, I think we will get some to carry a pregnancy."

All the cloning attempts, which date back to 2003, were filmed by Peter Williams, a distinguished documentary maker, for the Discovery Channel, which will show the programme tonight at 9pm.

Williams said that he was present at the secret laboratory when the cloning was carried out by Dr Illmensee. "There's never been any question of concealment, because we'd have known about it," Williams said.

The little girl who could 'live' again

Little Cady died aged 10 in a car crash in the US. Her blood cells were frozen and sent to Dr Zavos, who fused them with cow eggs to create cloned human-animal hybrid embryos.

These hybrid embryos were developed in the test tube and used to study the cloning process, but were not transferred into a human womb, despite Cady's mother saying she would sanction this if there was a chance the clone of her little girl could be born. Dr Zavos said he would never transfer hybrid animal clones into the human womb.

However, cells from Cady's "embryo" could in the future be extracted from the frozen hybrid embryo and fused with an empty human egg with its nucleus removed. This double cloning process could produce a human embryo that Dr Zavos said could be transferred into the womb to produce Cady's clone.

Frontiers of fertility: The key questions

Q. What does he claim to have done?

A. Panayiotis Zavos says he has created 14 human embryos and transferred 11 of them into the wombs of four women. Some of these embryos only developed to the four-cell stage before being transferred, but some developed to the 32-cell stage, called a morula. He also claims to have created human-bovine hybrid clones by transferring the cells of dead people into the empty eggs of cows. However, these hybrid embryos were used for research purposes and were not transferred to the womb.

Q. How does this compare to scientists' previous achievements?

A. Other scientists have created human-cloned embryos but not for the purposes of transferring them to wombs in order for women to give birth to babies. Those researchers created cloned human embryos in the test tube to extract stem cells for research. Dr Zavos has gone further (and broken a taboo) by creating embryos specifically for human reproduction, and he has attempted to create a viable pregnancy by transferring the cloned embryos into women.

Q. Hasn't he made similar claims before?

A. In 2004, Dr Zavos claimed to have transferred a cloned human embryo into a woman's womb but did not produce hard evidence. He has now produced more cloned human embryos, some at an advanced stage, and transferred them into the wombs of three more women. An independent documentary maker vouches for him.

Q. Why is this such a controversial thing to do?

A. Studies on animal cloning have shown time and time again that it is unsafe. The cloned animals suffer a higher-than-normal risk of severe developmental problems and the pregnancies often end in miscarriage. Mainstream scientists believe cloning is too dangerous to be used on humans.

Q. How likely is it that he will succeed?

A. He is determined to succeed and has a long line of people eager to sign up to his cloning programme, at a cost of between $45,000 and $75,000. Cloning attempts in other species, including primates, suggest there is no insuperable barrier to cloning humans."

Immortality has long been a dream of mankind. The successful religions have all had some "belief" or "faith" in some kind of life after death, heaven, paradise, etc. It has now become medically possible to successfully transfer cells from one human being to another. The only step missing is very close at hand. That is the transfer of the cells or group of cells that contain our memories. These memories when activated (awakened) may be the thing we refer to as a soul or simply "consciousness".

The successful experiment of Walter's memory transplant had very little if any ethical ramifications. You had one dead man and one man in a forever-persistent vegetative state and he was scheduled to have the plug pulled. The downside was nil. With cloning it is a more complicated matter. If you transplant your old memories into a new cloned you, are you destroying the virgin future of another person? If you were to do it and then ask the new "you" what would the new "you" say? Of course the answer would be, " Hell, I am glad to be back alive after I died! I also like this healthy young body quite well."

Here come the black helicopters!

Several months went by and rumors were out that real life after death had been achieved through memory transplantation. Although we knew it was the truth, the Government worked hard to make it all look like a hoax.

Each time someone would get close to the truth, Black Ops were sent in to falsely prove it was false! Eerily it was like the Roswell flying saucer incident all over again.

One day two men in black suits with dark sunglasses knocked on Walters door. They were from the US Department of Defense. As Walter answered the door they grabbed him and said, "You are under arrest for the act of treason!" "Damn" Walter thought, I will now be dissected for sure.

The two men put Walter into a black car with dark tinted windows. Walter asked, " Where are you taking me?" The main man said, "We are flying you to DC." When the car pulled up to the main gate at Mac Dill Air-force Base, Walter knew this was a big deal.

The flight to Washington on the Air force Hawker Jet was quick and the landing at Andrews Air Force Base was smooth. The next part of the trip was strange in that they took him to a small unobtrusive building on the airfield that had no windows. Inside he was guided to an elevator that went down instead of up. There were no floors listed and no elevator buttons, just a card reader and keypad. Down they went at a quick pace and Walter noticed his ears were feeling a slight pain from the increased air pressure. This meant that they had gone down into the earth to quite a depth. Being a pilot Walter knew it was at least a thousand feet down!

When they got out of the elevator they walked down a wide hallway and got into a small electric vehicle. This thing went very fast through a tiled lined tunnel. A few minutes later they got out and went into another room filled with all sorts of computer screens. Most were not on but some were lit up with numbers flashing across the screens.

A lady in a blue uniform asked Walter if he was hungry and he said no, but I would sure like a cup of coffee. She said, "I will be right back, cream or sugar?" Walter said, "Black." AS Walter sat there wondering what the hell had just happened to him, a familiar face appeared on one of the large monitors on the wall. It was the face of the President himself! The President said, "Mr. Gochen it is nice to see you again." I hope this has not been too traumatic. Let me assure you that you are not really under arrest or in any trouble, but we had to do this quickly and without public awareness. I will explain it all in just a little while; I am in an important meeting. I will meet with you ASAP." The screen went back to the presidential seal screen saver. The seal moved hypnotically around the screen. Walter thought, "What the hell, I am definitely on an interesting adventure and I might as well enjoy it." When the lady in the blue uniform brought Walter his coffee he just looked at it for a while and was scared to drink it. He imagined the operating table again! Autopsy was not a good thing!

About a half hour went by and the President with two men dressed in white smocks came into the room. The President said this is a very important meeting and we must come to a total understanding. We already have a secret board working on Lifespan rules and regulations. Now that we have proof beyond any doubt that near immortality is obtainable we must introduce it in a gradual way. If this gets out to the public at large we could have riots like we have never seen before. We will have to build many clinics and maybe have a national lotto to limit those who get in line to have their memory transplanted. It is simply impossible to make it available to all. Mr. Gochen, since you were the very first, we would like to include you in the plans and discussions. We will pay you well and take care of all your medical needs now and in the future. We do want to study your health in detail over the course of your second life. If you agree to all the terms we will put you at the top of the list for a third lifespan. During your second life span, should you die from an accident and your memory brain cells can be saved, you will have a government contract to re-transplant your memory cells into the next available donor giving you a third life span. I know this sounds a little like you being used as a guinea pig but you have absolutely nothing to lose. We only ask that you remain silent about any of our secret activities. Anyway if you don't agree we will have no choice but to arrest you for the sake of national security. With a shaky hand Walter signed a 17-page document. He thought, "What else can I do?"

Within a week Walter got his first check for $9000.00 dollars. This would be a monthly payment for this lifespan and be adjusted up or down for the next!

Plus all medical and travel expenses to and from Washington DC. Those Hawker jet trips from Mac Dill Field were nice and they even let Walter fly the jet for a few minutes!

About 20 years went by with not too much excitement for Walter. He basically had it made in the shade. Life was good, so far.

Cloning

The obvious next breakthrough would be to clone another you and transfer your memory brain cells to the younger clone. The basic scientific problem, which must be solved, is just when to transplant the memory cells. If the cells are transplanted at an early stage you may find yourself trapped into a small child's body for many years. This could cause severe mental problems. If you wait until the cloned child has developed it's own memory would you be killing someone by removing the existing memory cells and replacing them with yours? One solution would be to surgically remove the existing memory cells while in early development. The age-old question was, when did the clone begin to have human rights? It was decided by the government Life-Span board to use the legal abortion age. If the memory cells were removed from the clone while still younger than the legal abortion age, it would be legal to remove the memory cells. The cloned child would be vegetative until the memory cell transplant was done. This meant that you had the trade off of keeping the clone alive until older, versus the cost of maintaining the clone. The younger the clone the lower the total cost. The transplant operation cost was cheap compared to maintaining a clone in a vegetative state year after year. I feel sure that when memory transplants becomes public, Wall Street will come up with an insurance plan to pay for it. Some might say immortality is a bargain at any cost.

Although any part of the human body can now be cloned successfully, ethical and religious questions still are being wrestled with. Complete livers, spleens, kidneys, hearts, and even lungs are routinely cloned and transplanted with no immune rejection. Since they are made with the exact DNA in each cell they are perfect matches with no immune rejection side effects.

Another Trip to Washington DC

Walter was brought to Washington DC for the purpose of revealing to him the latest advancements in cloning. He was told that cloning a complete human body has only been done in top-secret labs. One of these labs is hidden away 1000 feet below Andrews Air Force base in Washington DC. Inside the lab is a room filled with large jars with placentas that are connected to a device similar to a dialysis machine. Blood is pumped through the placentas and feeds the developing fetuses. The room itself is temperature controlled to exactly 98.6 degrees F. This room is very uncomfortable but is kept at this temperature as a failsafe and to simulate a mother's womb temperature. Each fetus jar had a brass plate attached near the bottom. All had the same format "Name of Clone Donor-Second Life-Span" Walter found a jar with a nameplate that said "Al Gochen Third Life-Span". He could only conclude that the Feds were planning on later transplanting his memories into his clone. It looked like they were going to honor the promise of not only a third

Life Span but also transplant his memories into a clone of himself!

As Walter walked along the row of jars he noticed that one was different from the rest. He looked at the nameplate and thought, "Now I understand why all the secrecy about his birthplace and education".

The brass nameplate read:

BARACK HUSSEIN OBAMA—FOURTH LIFESPAN.

The End

Footnote: Most of the medical and scientific elements in this book are true but we have not found the location of our memories yet, or how to successfully transplant the brain cells that contain them.

Or have we?

The news article is real. Check on Google: Dr. Zavos.